Much Math Fun

BY KIM THOMPSON

A Little Honey Book

Crabtree Publishing
crabtreebooks.com

Tips for Teachers and Caregivers

This book supports early readers as they decode words to learn facts and gain knowledge about the world.

Before reading, make sure students understand the sound-spelling correspondences shown below as well as the high-frequency words shown on the next page. Introduce the vocabulary words.

During reading, provide feedback and encouragement as students sound out decodable words by blending individual sounds.

After reading, talk about and write about the topic. Share the information on page 16 to help students learn more.

Letters and Sounds

New:

consonant digraphs *ch* and *th* (voiceless)

Review:

all consonant sounds and standard spellings; consonant digraphs *ng* and *sh*; *short a* spelled *a, short e* spelled *e, short i* spelled *i, short o* spelled *o, short u* spelled *u*

Decodable Words

an, branch, bunch, champ, check, chest, chick, chimps, chin, chums, clutch, crunch, depth, eggs, end, fifth, fish, hands, has, hatch, help, in, inch, is, it, itch, length, lips, lunch, match, math, much, munch, patch, path, sixth, snatch, such, thick, thin, think, which, width, will, with

High-Frequency Words

New: each, most, soon, try

Review: a, all, are, be, by, do, for, go, have, how, long, many, more, of, one, or, see, take, the, they, this, to, walk, wants, white, you

Vocabulary Words

arms

caterpillars

seals

This bunch wants to munch.

Match each with a lunch.

Which lunch has much to crunch?

Check the clutch of eggs.

Which chick will hatch soon?

Is it the fifth or the sixth?

More Math...

4+______=6

1
2
3
4
5
6

Do you see thick lips?

Do you see a long, thin chin?

Do you see a chest with a patch of white?

More Math...
How many in all?

Caterpillars go inch by inch.

They take a path to the end of each branch.

Which will walk the most length?

caterpillar

The chimps are
such chums.

They help with an itch.

Which **arms** have the most width?

More Math...

How many chimp hands in all?

Seals try to snatch a fish.

Do you think one will be the champ?

Which has the most depth?

More Math...

How many fish for each seal?

seals

Build Background Knowledge

It is fun to count animals, notice things about them, compare them, and classify them into categories. All these skills are important for building a strong foundation in math. After reading the book and thinking about each concept presented, find more ways to use numbers with the animals. Can you make up story problems about them? Can you sort them into different groups and count the number in each group?

MUCH MATH FUN

Written by: Kim Thompson
Designed by: Rhea Magaro
Series Development: James Earley
Educational Consultant: Marie Lemke, M.Ed.

Photographs: All images from Shutterstock

Crabtree Publishing

crabtreebooks.com 800-387-7650

Printed in China/012024/FE20231222

Published in Canada
Crabtree Publishing
616 Welland Ave.
St. Catharines, Ontario
L2M 5V6

Published in the United States
Crabtree Publishing
347 Fifth Ave
Suite 1402-145
New York, NY 10016

Library and Archives Canada Cataloguing in Publication
Available at Library and Archives Canada

Library of Congress Cataloging-in-Publication Data
Available at the Library of Congress

Hardcover: 978-1-0398-4441-4
Paperback: 978-1-0398-4522-0
Ebook (pdf): 978-1-0398-4599-2
Epub: 978-1-0398-4669-2
Read-Along: 978-1-0398-4739-2
Audio: 978-1-0398-4809-2